The Alpaca Breeding Book

Alpaca Reproduction
&
Behavior

by
Dr. KD Galbraith

Walnut Creek Publishing
Tuskahoma, Oklahoma

The Alpaca Breeding Book

ISBN: 978-0-9893241-0-6

Acknowledgments

Many thanks to our fellow alpaca breeders for their help over the years as we learned about breeding alpacas. A big thank you to our local veterinarians and Oklahoma State University veterinarians for their help and research.

Thank you to our partners for their love and understanding. Also, thank you to my mom, Theresa DiVita, and my family for their love and many prayers. Above all, my husband David Galbraith, who continues to provide wisdom, stability, and much love that enables me the opportunity to express myself and write.

Dave with a newborn cria

Two crias playing and practicing

Table of Contents

Introduction

The Alpaca Breeding Book will help alpaca owners learn about alpaca behavior and how they reproduce. Alpacas are "induced ovulators" so they are much different from other livestock. Every breeding season can be a challenge if you are not familiar with the way alpaca's breed. This breeding book, used as a tool, will help guide the breeder through different factors that can affect the female alpaca's ability to get pregnant.

With over 20 years experience breeding other livestock, it was a surprise to find out alpacas reproduce much differently. When we started raising alpacas there was little information about their breeding practices so we went through some challenging times. When we had trouble getting a female pregnant we would search the web, call our veterinarian, call Oklahoma State University, and go through my medical books to find the answers to our questions.

The first time we bred a female alpaca we brought an older male to the female and he bred her for about an hour. The breeding was so different from anything we previously raised. My husband and I looked at each other in amazement. We had nothing to compare the breeding to so we wondered if this was normal. The Alpaca Breeding Book will help the breeder have a better understanding about all aspects of breeding alpacas.

Not every female is an easy breeder as you may already know. Let's not blame everything on the girls since there are some boys not getting the job done as well. We cover female and male behavior and how their behavior, diet, and other factors can affect their ability to reproduce.

Mother alpaca with her first cria

Please note we are not veterinarians, however we have over 12 years experience breeding hundreds of alpacas. The information contained in this book comes from our experience and our veterinarian's advice, as well as input from other breeders. We communicate and share information with alpaca farms around the world.

In this book we share what works for us, using responsible, healthy breeding practices. Nine times out of ten, you won't experience any problems. When you need help, I know "The Alpaca Breeding Book" will be a great reference tool for anyone raising and breeding alpacas.

Please note the information in this book should not replace the expertise of a licensed veterinarian. This book will serve as a guide as you breed your alpacas and it will help you get your females pregnant next breeding season. After reading "The Alpaca Breeding Book", I wish you many successful breeding seasons and a brighter future raising alpacas.

1 - Breeding Introduction

Basic Breeding Questions

What is a good age to start breeding?

Female alpacas start to breed around 18 months of age if they are ready. We prefer the young girls to be at least 100 pounds in weight. The males start breeding effectively around 2.5 years old but this can vary and will be discussed in depth later.

How long are female alpacas productive?

If the female alpaca is in excellent condition she can continue to reproduce at 20 years and older. The oldest female we know of to continue to reproduce was 22 years old.

How are alpacas different from other livestock?

Alpacas are "induced ovulators" meaning the female ovulates after breeding with the male. However, ovulation only occurs when a mature follicle is present on the ovary. Also note, open females are almost always receptive to the males.

How long does the breeding last?

Most males will actively breed the female around 20 minutes. However, a male can breed anywhere from 5 to 60 minutes. The male will orgle (sing) the entire time he is breeding the female.

What happens when alpacas breed?

The breeding male will start to orgle (sing) as he approaches the female. The nonpregnant female alpaca will cush and she will let the male mount her. The nonpregnant female goes into a trance-like state while the male is breeding and singing to her. On the other hand, a pregnant female will spit or run from the male. Please note that each female may react a little differently when the male alpaca approaches.

Young male crias practice breeding

How long is an alpaca's gestation?

The female alpaca has a gestation period between 335 to 365 days but it is not uncommon to go past 365 days. Our gestation calculator online is set to a 340 day gestation. You are welcome to use our gestation calculator online to see when your female is due. Please refer to the resource section for the web address.

Can alpacas have twins?

Yes, they can have twins but it is uncommon and they rarely carry twins full term. See Goldie's story in Chapter 2 under Twins.

Is there a breeding season?

They do not have a breeding season. We choose the time of year when we want them to breed. We breed the alpacas when the weather is appropriate for having crias. In our part of the country, breeding for the spring season is April through May and the fall season is October through November.

What are the breeding methods?

There are two different methods of breeding, hand breeding and pasture or field breeding. We will concentrate on hand breeding in this book. Hand breeding is when the male and female alpacas breed in a controlled environment. Field breeding is where the male and female remain together in the pasture without supervision.

Can males run with the females?

Pasture breeding is not uncommon but we recommend hand breeding the female to gain control over timing of birth. Since alpacas are induced ovulators, an intact male should not remain the entire year with the females. The male will breed any open females and they will have crias at different times of the year. Many crias are lost due to birthing in cold weather. We will explain the different breeding methods in further detail in chapter 5.

Alpacas breed in a cushed position

Induced Ovulation

How are alpacas different from other livestock?

Most livestock have an estrous cycle. One example is cattle. Cows have an estrous cycle, which gives a cow the chance to become pregnant every 21 days. They also have an anestrous period where they do not cycle and they are not receptive. They cannot become pregnant during the anestrous period.

What is induced ovulation?

Alpacas are different because they are induced ovulators. Alpacas do not have an estrous cycle as other livestock. Induced ovulation is much different because ovulation occurs around 24-48 hours after breeding. They are usually receptive when a follicle is present. A follicle is present most of the time because as one follicle is growing on one ovary another follicle is regressing on the other ovary.

A follicular wave is the cycle of the follicles growing and regressing. The female alpaca's follicular waves overlap. There are wave charts on the Internet that calculate the best time to breed the female. The wave charts never worked for us. The reason they may not work well is because all females are not the same. We have better results watching our female's behavior rather than sticking to a chart with a specific day to breed.

When the alpacas breed the male alpaca orgles (sings) to the female the entire time. The female will space out and some will go into a trance-like state. Through the act of mating the female receives signals to her brain. As a result, the female alpaca releases hormones in her body.

The hormones released are Gonadotropin Releasing Hormone (GnRH), Luteinising Hormone (LH) and Follicle Stimulating Hormone (FSH).

Follicle Stimulating Hormone (FSH) stimulates the follicles to develop. Luteinising Hormone (LH) stimulates the ovary to release the ova (ovulate) from a mature follicle. The mature follicle ruptures and releases an egg. In 3 to 6 days, a Corpus Luteum (CL) forms on the ovary. The Corpus Luteum (CL) produces progesterone, which makes the female nonreceptive. The female produces progesterone to maintain pregnancy if fertilization occurs.

It takes 5 to 7 days for the released egg to travel towards the uterus. If the egg is fertilized, it will take around 3 to 4 weeks for the egg to attach to the uterus. Most of the time, the egg will attach on the left uterine horn. This is helpful to know when performing ultrasound pregnancy testing.

If fertilization does not occur, the Corpus Luteum (CL) will regress in about 10 to 13 days. The progesterone levels will decrease and the female will become receptive again.

Note:

This is an overview of the female alpaca's reproductive physiology and an introduction to induced ovulation. By keeping the alpaca's breeding process simple, you will have a better understanding of the basics.

For an in depth look at the alpaca's anatomy and physiology, refer to references and further reading. Key Reproductive Features, by the Australian Alpaca Association, is an excellent online resource and is available as a download in pdf form.

Factors Affecting Reproduction

If the female alpaca is having trouble getting pregnant, there are several factors to consider:

- ➢ Age
- ➢ Behavior
- ➢ Health
- ➢ Hormones
- ➢ Nutrition
- ➢ Physical
- ➢ Stress
- ➢ Weather
- ➢ Weight
- ➢ Wormers/ Medications

We will discuss each factor listed in depth. We will cover the females first and then tackle the males since it does take two to tango!

2 - Female Alpacas

Breeding Age ~ Females

Young females, maidens, can start breeding at 12 months of age but we feel there are fewer problems if we wait at least 6 months longer before breeding. We prefer to start breeding our maidens around 18 months of age. As a rule, we also prefer their weight to be at least 100 pounds. Please note there may be a few maidens not ready to breed even by age two.

Reasons to wait until 18 months of age:
> ➢ Not physically mature - slow growing females, small in stature.
> ➢ Not mentally mature - females afraid of males.
> ➢ Fewer problems with dystocia – abnormal or difficult childbirth or labor.

If a maiden acts fearful towards the male, we try different ways to keep her calm. We use a less aggressive male preferably one the female is attracted to. Some females prefer certain males and they will flirt with them over the fence or from a distance. The alpacas are smart and each one has their own unique personality. The alpaca's behavior can affect the breeding.

Young females flirt and tease the male

Note:

When spit testing a young female, use a less aggressive male since a young female may cush for an aggressive male out of fear.

We breed a maiden only three times in a breeding season. If she fails to become pregnant, we wait and try again the following breeding season. Sometimes waiting a breeding season or two can make a big difference.

Behavior ~ Females

Behavior or mental side of breeding can easily affect reproduction, especially since alpacas are induced ovulators. Ovulation only occurs when the female is bred with a mature follicle present. When the female breeds the brain releases chemicals. Thus fear, like and dislike of a certain male can affect the female alpacas ability to become pregnant.

Attraction

A few females may not like the male you selected for them. In fact, they might not want anything to do with him.

Our solution is to bring another male to her. Find a male that makes her fall to her knees at the sight of him. When she cushes (do not give him a chance to do anything) pull him off and bring in the male you want her to breed to. Most of the time, she will stay cushed and breed with the other male you picked for her.

Difficult females

Some females will run and spit when they are open. One way to know if they are open or pregnant is to let the male mount the female.

We have a female that spits and runs from the male every time. I have to put her in a smaller pen and tell the male, "her no means yes so don't give up too quickly". Although, as soon as the male touches her back, she falls to her knees, cushes, and allows the male to breed her.

I know what you're thinking… how do I spit test a female that spits when she is open and how will I know if she's pregnant?

One way to test difficult females is to test them in a small pen. If the girls are in a large pen they will run away from the male before he has a chance to do anything. With our difficult female when she's open, she runs away spitting but when she's pregnant, her eyes open wide and she gets frantic. She runs away spitting as if to say "not just no, but hell no!!!"

Young female afraid of males

Fear

Maidens afraid of males

When a young female is afraid to breed, bring a calmer experienced female into a small pen with another male. Breed the older female first so the young female can see there is nothing to be afraid of. She may cush next to the breeding pair. Next, bring the young female's mate back into the pen. If she will stay cushed, let the male breed her next to the other breeding pair. If the female still refuses to cush and breed, it is usually best to wait a breeding season until she is more emotionally mature.

Personality

Alpacas are intelligent animals. It is better to know ahead of time that you are dealing with personalities not just livestock. Every alpaca has a different personality and each one behaves differently open versus pregnant.

Open females have a sweeter personality and they usually do not mind being handled. Most pregnant females act cranky and do not want to be touched. The bred female's personality change is because of the rise of progesterone in her body. Many alpaca owners will say, "What happened to my sweet female?" "She doesn't like to be handled and she spits at the drop of a hat". If she is persistent in not wanting to be handled, she is probably pregnant.

When spit testing, each female alpaca will react differently when you bring a male into their pen. If you know the alpaca's personality and behavior in advance, spit testing becomes a useful and accurate tool for pregnancy testing.

Group behavior

Alpacas are herd animals so most females will follow the group. They are also flight animals so they are different from other livestock. By instinct when one female runs away, they all run away. When alpacas are excited they run with the herd for survival.

At times when I walk through a pen of open females with a male, instead of being receptive, they will run and spit at the male. To slow everything down I put a few open females in a smaller pen. When I bring the male to them, they can't run away so it is easier for the male to convince them to breed with him. Once the male starts to breed a female, the open females hear the male orgling and they will cush by the breeding pair.

Several open females cush next to the breeding pair

By allowing open females to see and hear other females breeding it will help them to start cycling. Breed the cushed females as soon as possible, preferably the same day. Once their progesterone rises the breeder will have to wait at least one to two weeks before the female will become receptive again. Some females will continue to cush for as long as week after being bred. Other females start to spit shortly after being bred. It is helpful to remember, every female alpaca is different.

Health Problems ~ Females

Two common health problems associated with reproduction are Uterine Infections (UI) and Retained Corpus Luteum (CL).

Uterine Infection (UI)

With over twelve years experience of breeding hundreds of alpacas, we only had one female with a uterine infection. She came from another farm to breed one of our males but she never became pregnant. The female was receptive around every 14 days. She never had discharge or any other symptoms. The female was visiting our farm so we did not know what breeding practices she was exposed to in the past. We brought her to OSU and they confirmed she had a uterine infection. They flushed out her uterus and gave her antibiotics. When her treatment was over, she became pregnant after one breeding. The following year she gave birth to a healthy cria.

Through research and discussion with veterinarians, we discovered the common cause of uterine infections was over-breeding. In an article written by Tibary & Anouassi (2001) on Uterine Infections they state "the most significant factors contributing to uterine infections in camelids are over-breeding (i.e. excessive matings during the period of receptivity), postpartum complications and unhygienic gynecological examination and manipulation... the average number of matings per breeding season (generally a period of 20 to 40 days) before referral was 8.5, with a between mating interval of 3.2 days."

After reading the article on Uterine Infections, there is a problem with some alpaca breeders over-breeding their females.

Please use responsible breeding protocols because breeding more than once a week will do more harm than good. At our farm, we stick to our 7-day schedule and never breed more than once a week.

Treatment for a uterine infection includes antibiotics, uterine flush with a small catheter, and wait at least 2 weeks before trying to breed the female again.

Retained Corpus Luteum (CL)

We had one case of a retained Corpus Luteum (CL) on our farm. In fact, it turned out she had two retained CL's. The female's odd behavior was her only symptom. At times the female acted as if she wanted to breed but when the male approached her, she would spit and run away from the male. The retained CL's were producing progesterone so she acted like a pregnant female most of the time. There were no other symptoms.

We brought her to Oklahoma State University and they found two retained corpus luteums by performing an ultrasound. They checked her progesterone and it was elevated. They gave her a prescription for Estrumate (cloprostenol sodium) at a normal dosage and normal frequency.

After her prescribed treatment, the female continued to spit off the males and she continued to have a high progesterone level. The doctors at OSU increased her dosage and frequency not once but twice. Finally, the female's progesterone levels dropped and she was receptive to the male again. After breeding her to the male she became pregnant easily and she gave birth to a normal, healthy cria the following year.

Hormones ~ Females

Progesterone

Progesterone plays an important role in pregnancy. If there is a progesterone deficiency, the female will be unable to maintain her pregnancy. A progesterone deficiency is treated with hormone injections so the female can carry her fetus full term.

The following is a typical scenario where we might suspect a progesterone deficiency; the female is bred and she spits at the male on day 7 and day 14. Sometimes she will continue to spit at day 21. Around day 30 she is receptive and cushing for the male. The female is open again and wanting to breed.

Several farms we spoke with had females with a progesterone deficiency. It is interesting to note the breeders from these farms had similar stories... once the progesterone deficient female had her first cria she went on to have normal pregnancies. Hormone injections were no longer needed for the next pregnancy and the female's progesterone levels remained within normal range.

Gonadotropin Releasing Hormone (GnRH)

Gonadotropin Releasing Hormone (GnRH) is used to induce ovulation. GnRH is normally secreted by the hypothalamus and it acts on the pituitary gland to release Luteinising Hormone (LH). The LH acts on the follicle and ovulation occurs. GnRH is by prescription through a veterinarian.

We brought a female to Oklahoma State University that would breed readily but she never became pregnant. They did a full reproductive examination on her. The doctor's could not find anything wrong with her. They said it appeared she was not ovulating so they decided to treat her with a Gonadotropin Releasing Hormone called Cystorelin. The doctors instructed us to give her the Cystorelin and then breed her 24 to 36 hours after the injection. We followed their instructions but she never became pregnant. After several attempts treating and breeding, we decided to remove her from our breeding program. Sometimes it is better to cut losses and move on.

Estrumate (cloprostenol sodium)

The label states "An Analogue of Prostaglandin F2 for Intramuscular injection. Equivalent to 250 mcg cloprostenol/ mL. The luteolytic action of Estrumate can be utilized to manipulate the estrous cycle to better fit certain management practices, to terminate pregnancies resulting from mismatings, and to treat certain conditions associated with prolonged luteal function."

Estrumate is by prescription through a veterinarian. There are warnings on the label because it can be absorbed through the skin. Handle all hormones with extreme care.

Estrumate is used for the following:

➢ Manipulate female's cycle
➢ Terminate pregnancy
➢ Induce labor
➢ Retained Corpus Luteum(CL)

Your veterinarian may prescribe the Estrumate differently than we describe. The following information is instructions from our veterinarian. We will look at the different uses of Estrumate in further detail.

Manipulate female's cycle

Alpacas are induced ovulators but some females may need help getting started. They may refuse to breed even though they are not pregnant. The Estrumate will help the female to start cycling. We inject the female with 1cc of Estrumate in the muscle. The female should become receptive 3 to 10 days after the Estrumate is given. If the female is not receptive by the 10th day, we repeat the Estrumate injection. She should be receptive and willing to breed by the 3rd or 4th day.

Terminate Pregnancy

Estrumate is given to a pregnant female to abort an unwanted pregnancy from an unplanned or wrong breeding. To terminate pregnancy we give Estrumate twice, either 2 consecutive days or on day one and again on day three. In one week, we check the female with a male to see if she is receptive. We give a second dose of Estrumate if she continues to spit off the male. Then we check her again in one week and she should be receptive and cush for the male.

Induce Labor

There were times when we needed to induce labor and the Estrumate proved to be helpful on several occasions. When we give an Estrumate injection to induce labor, the female will usually deliver her cria around 24 hours later.

Please consult with a veterinarian and be very cautious when inducing labor in an alpaca because the gestation time can vary greatly from one female to another. The gestation time can be anywhere from 330 to 365 days or longer.

Reasons we induce labor:

➢ History of going past the due date and the female or cria has problems

➢ History of difficult labor because of large crias

➢ History of having crias with crooked or swept legs

Several of our females, especially when they are bred for a spring cria, go over 365 days. These females end up with more problems because of the longer gestation period.

Newborn cria

One female's crias were too big for her to deliver. She had difficult births and she would lose her crias. When we give her Estrumate around day 350 she has easier births and live crias. Another female we had would have crias with crooked legs. We now give her Estrumate around day 350 and now all of her crias are born with straight legs.

It is important to discuss the options with your veterinarian so you can make an informed and responsible decision about inducing labor in your alpaca.

Retained Corpus Luteum

A Retained Corpus Luteum (CL) is a condition where the Corpus Luteum (CL) persists and causes the female to have elevated progesterone levels. The female is not receptive to the male. We give Estrumate to treat the retained corpus luteum. The Estrumate will cause the CL to regress. In turn, the progesterone should return to normal levels. The female will start to cycle and she will become receptive again.

For more information on Retained Corpus Luteum (CL), see the Health Problems, section under Retained Corpus Luteum.

Physical ~ Females

Persistent Hymen

At times, a female may fail to conceive because something physically is stopping conception. A persistent hymen is a possible cause often overlooked.

Some breeders have their veterinarian do a reproductive check-up on their maidens. During the check-up, they have their veterinarian break the maiden's hymen. Other alpaca breeders we spoke with leave it up to the male to break the hymen and let nature take its course.

We prefer to let the male break the hymen when he breeds the maiden. Over the years, we never had a problem. If you suspect a persistent hymen and the male has not been able to penetrate the female, contact your veterinarian. The female alpaca's hymen may have to be broken down manually.

Note:

We check the female's genitalia before and after the breeding to see if there is discharge. After breeding, the female may expel a little blood or semen and this is normal.

Uterine Prolapse

With a uterine prolapse, the female's uterus is expelled after she gives birth. A uterine prolapse is usually associated with a difficult birth or intense pushing.

I was present when two of our females had a uterine prolapse. Neither of the females had a difficult birth. In fact, both females had easy deliveries and passed their placenta with ease as well.

Putting the uterus back in place after a prolapse is difficult and should be left to the veterinarian. When we find a female with a uterine prolapse, we rinse off the uterus with water to keep it moist and to clean off the vegetable matter. Then we wrap the uterus with a large plastic garbage bag. Try to keep the female calm and take care not to puncture or tear the uterine wall. Proceed immediately to a veterinarian.

After the veterinarian puts the uterus back in place, he puts a few stitches around the vulva in order to keep the uterus from coming back out. The female is given antibiotics to prevent infection. Sometimes oxytocin is given to facilitate contractions in order to reduce the uterine swelling.

Some breeders have a misunderstanding about a uterine prolapse. They believe if a female has a uterine prolapse, she can no longer breed. Please note that after a uterine prolapse a female can go on to breed and deliver normally with no problems. We had three females over the years have a uterine prolapse, all three females never prolapsed again, and they went on to have normal healthy crias.

Note:

After speaking with our veterinarian we found out if they have a vaginal prolapse, there is a higher risk of prolapsing again. We were fortunate with our three females because they never had any problems associated with their uterine prolapse.

Female cria spits at male cria

Stress ~ Females

Alpacas have a lower early conception rate compared to other livestock. The lower early conception rate may be related to stress. Stressed females can have a difficult time getting pregnant or have trouble staying pregnant. Stress can play a big part in breeder's losses without the breeder realizing what is happening.

The following conditions can cause stress:

➢ During breeding – hand or pasture breeding

➢ Pecking order

➢ Transporting

➢ Unknown sources

We will discuss the possible stress conditions mentioned above in further detail.

During Breeding

Hand Breeding

With hand breeding, keep the distractions to a minimum by not interfering with the breeding. As previously mentioned hand breeding is when the male and female alpaca breed in a controlled environment.

If hand breeding is the method of choice for breeding alpacas it is best to stand back and let them breed. If the tail is not wrapped it is ok to move the female's tail out of the way. Let the female go into her trance so her brain will release the hormones needed so she will be able to ovulate. Try to keep the human interference to a minimum.

Pasture Breeding

Field or pasture breeding is where the male and female remain together in the pasture without supervision.

When pasture breeding, be sure to remove the male from the female's pen before the last two months of pregnancy. The male can stress the female by continually checking her to see if she wants to breed. Another concern as the female gets close to her due date her progesterone levels start to drop. She may let the male mount her and breed her possibly causing damage to her or the baby. It is best to remove the male from the female's pen shortly after breeding season.

Alpacas have a pecking order

Pecking order

Alpacas have a pecking order. When they are comfortable with their place in the pecking order, there is peace in the herd. However, some females may stress other females in the herd by picking on them.

If a female higher up in the pecking order picks on a female lower in the pecking order, the situation could be stressful for the lower female. In addition, if the pens are overcrowded, the pecking order can become more intense and this situation can easily occur without your knowledge.

We try to pen alpacas according to age and size. They have less stress and they are happier with other alpacas in their own age group. Grouping by age groups would apply more to a breeder with a larger herd.

Transport

Transporting a female in the first two or last two months of pregnancy is too stressful for a bred female. Do not transport a bred female during those critical months.

When transporting a pregnant female we usually give a shot of Banamine. Banamine is an anti-inflammatory that helps the pregnant female during transport. Banamine needs a prescription through a licensed veterinarian.

Try to avoid transporting a pregnant female by herself. Make sure the female has a companion, other alpacas, to travel with.

Unknown sources

Stress can come from something outside the herd. It can be as simple as the neighbor's dog or your dog. Sit back and watch your herd and they will tell you what is stressing them. Even something simple can stress a female into not breeding or cause resorption and loss of pregnancy.

Keeping an alpaca alone without the company of another alpaca, can cause enough stress to cause resorption of the embryo or abort. Alpacas are herd animals and they need the companionship of other alpacas.

Twins ~ Females

Alpacas rarely carry twins full term. It is possible for the female to abort one twin and carry the other twin full term. The following is Goldie's story.

Goldie's story

One year we bred our female, Goldie, in the fall. After one breeding, she became pregnant. The following season in the spring, she aborted a female fetus. I tried to breed her back but she was not interested in the males. Every time I brought a male to her, she would spit and run away from the male. I decided to leave her open and breed her back in the fall.

Later the same year, in the fall, I brought a male to Goldie and again she spit and ran away from the male. I knew we had to investigate further and find out why she did not want to breed.

Alpaca twins are rare

Before we had the chance to do anything, Goldie delivered a cria. It was shocking because she did not look pregnant. She gave birth to a beautiful healthy female cria. What a surprise, we never thought she could still be pregnant after aborting in the spring.

Note:

When a female alpaca does not want to breed, pregnancy should be considered as one of the possible causes.

Freemartin

Alpacas rarely have twins but be aware of what could happen if an alpaca has twins of the opposite sex. The female will be sterile. She will have non-functioning ovaries, which is a normal result of opposite-sex twins. The female becomes sterile in the womb by hormones from the male twin. She is a freemartin and cannot reproduce.

We were fortunate that Goldie's twins were both female so her cria was not affected by this in any way.

Weather ~ Females

Weather and seasonal influence has affected our breeding and birthing in the past.

Heat Stress

Females may abort when exposed to high temperatures for prolonged periods. The heat affects each pregnant female differently and it may depend on how far along she is in her pregnancy.

Excessive heat during the first two months of pregnancy can cause fetal resorption of the embryo or defects. Exposing a pregnant female to excessive heat at the end of pregnancy may cause her to abort or have a premature cria.

Tents provide shade to help keep the alpacas cool

Our bred females experienced many different hardships during a heat spell where the temperature was well over 100 degrees and lasted two full months. Several farms had an increase in stillbirths plus defects we believe were caused by the prolonged intense heat.

On our farm, we had two crias born partially blind. The mothers of the crias conceived within a week of each other so they were both at the same stage of pregnancy when we had the high temperatures.

Sienna stays cool in the kiddie pool
Don't let them stay in the pool too long their fleece will rot!

Note:

To see how to keep alpacas cool in the summer look at chapter 3 under the male's weather section titled "Beat the heat". There are great ideas and important information to help keep the alpacas cool during the hot summer months.

Strange weather patterns

Odd weather can interfere with breeding effectiveness. One year our spring was unusually cold, dark, and rainy. The females did not want to breed. Even though alpacas are induced ovulators the females refused to breed regardless of what we did that spring. Eventually the sun came out and the alpacas were finally wanting to breed again. It was apparent the amount of daylight affected their willingness to breed that spring.

We called Oklahoma State University and they stated that across the board most breeders, not just alpaca breeders, were having a difficult time getting their animals to cycle and start breeding.

Spring versus fall crias

In our part of the country, we noticed our females due in the spring, usually deliver later and their crias need more help getting started.

Crias born in the fall were usually on average about 2 weeks early. The fall crias have more energy and they did not need as much assistance as the spring crias.

The spring versus fall crias varies greatly on our farm. We now breed predominately for fall crias. The difference in crias may be due to the various weather patterns or different forages available in the pastures in the spring as compared to the fall.

Weight ~ Females

Obviously, we should only breed healthy alpacas. An alpaca's weight can affect their ability to reproduce. Alpacas can have trouble getting pregnant if they are too thin or fat. They may also have trouble maintaining their pregnancy. Try to make sure the breeding females are in excellent condition before breeding them.

Thin females

Some females have trouble keeping the weight on when they are nursing a cria. They can get thin quick and they can have trouble getting pregnant. Adding milk plus or calf manna to their feed helps keep the weight on. If we have a thin female with a nursing cria that is not getting pregnant, we wean the cria first and then breed her again the following breeding season.

Including a soft mineral block as free choice helps nursing moms stay healthier. Make sure to use a sheep mineral block because it is low in copper. Our alpacas eat the soft mineral block much better than the loose minerals.

Overweight females

Heavy females sometimes have trouble with breeding and birthing. Some overweight females have trouble maintaining a pregnancy. They may be unable to carry full term and have premature crias. Another problem associated with obesity is a lack of milk for their cria.

A simple way to monitor an alpaca's weight

Place your hand on the alpaca's back while they are eating at the trough. By doing this regularly when you feed, it acts twofold, they will become familiar to your touch as well as monitoring their weight.

When you place your hand on the alpaca's back, you should feel the backbone. The backbone will form a tent on the alpaca's back. If the tent is sunken down on the sides, the female is too thin. If the back is round and no backbone is found, the female is too fat. In addition, an overweight alpaca retains fat in the chest area. If a female's chest jiggles when she walks, she's probably a little overweight.

Note:

An alpaca's fleece can hide weight loss. An alpaca can lose many pounds before you realize what is happening. Not everyone has access to scales. We didn't have scales when we first started raising alpacas. We had to find a quick and easy way to monitor our alpaca's weight without scales. Placing your hand on their back while they are eating at the trough is a quick and easy way to monitor an alpaca's weight.

Wormers/ Medications ~ Females

Different wormers and medications may interfere with reproduction. Try to avoid using wormers and medications especially the first two months after conception.

Information on alpacas and their reactions to drugs is scarce. Wormers and medications have no information on their label about dosages or product safety for alpacas.

Over time, by either personal experience or from other farm's experiences, we gather information on the different drugs and how they can affect an alpaca's pregnancy.

Reactions

➢ Never give more than one medication at a time. Anaphylactic shock is common when multiple medications are administered.

➢ Never give pregnant females corticosteroids. Minute amounts in the eyes can cause abortion.

➢ Some wormers can induce abortion such as the wormer Valbazen.

➢ Use Estrumate not Lutalyse. Our understanding is Lutalyse can kill an alpaca.

➢ 7-way or 8-way should not harm a pregnant female or the unborn fetus. However, we know several farms that used 7 or 8-way on pregnant females and the females aborted.

The drugs and wormers listed should never be used on pregnant females. We rather error on the side of caution than take a chance on our females aborting or cause harm to the unborn fetus.

Precautions

- ➢ Never use chemicals on alpaca pastures.
- ➢ Never put an alpaca on a field after using a weed killer. Do not use weed killer even if it states it is safe for other livestock to graze on after using.

Note:

We know several farms that lost alpacas after grazing on pastures 5 weeks after treating the pasture with 2-4D weed killer, which is safe for most other livestock.

Alternative ideas to control worms

- ➢ Rotate pastures
- ➢ Avoid overcrowding
- ➢ Pastures and paddocks should have good drainage
- ➢ Wait until noon before allowing alpacas on pasture. By waiting, the wet grass has a chance to dry.
- ➢ Most worms are host specific so let a horse graze on the alpaca pasture every once in awhile.

Injectable Ivermectin

We live in an area where there is meningeal worm present. We give injectable Ivermectin to all the alpacas monthly during the rainy season regardless of where they are in their pregnancy. So far, we have not seen any ill effects using the injectable Ivermectin during the first two months or last two months of pregnancy.

Note:

An excellent reference book on dosages for wormers and medications is Dr. Norm Evans, Alpaca Field Manual... for more information refer to the references and further reading section.

Mom and cria in pasture

3 - Male Alpacas

Breeding Age ~ Males

Young males

Most males become effective breeders between 2 to 3 years old. However, some males can get a female pregnant before they reach two years old. We take the males out of the female's pen and wean them around 6 months of age. We never leave a young male in with the females past 7 months of age. Most likely, they will not get a female pregnant but we have heard of farms having young males less than 12 months of age impregnating a female. Accidents can happen so it is better to remove the young males from the female's pen before they reach 8 months of age.

A young male alpaca may not breed successfully because the prepuce restricts the penis from fully extending. In young males, the prepuce is attached to the penis and this is normal in immature males. The adhesions will no longer be present when the male reaches sexual maturity between 2 to 3 years old. Another thing to watch for around this age, the male's fighting teeth come in. It is a good idea to check the males in this age group and cut their fighting teeth to avoid accidents.

When breeding a young inexperienced male the first few times, try to find easy females for him. Older, dominant females can intimidate young males so it is best to find a female that will cush easily and not spit at him. We want to make his first breeding rewarding... what a great way to start his breeding career! The next time we take him to the females he will be eager to breed.

Slow to mature

Some males mature slower than others. By age three, most males will be effective producers. However, do not count them out if they are not producing by age three. We had a male that was not productive until age five. As it turned out, he is one of our top producers and well worth the wait.

Older males

Older males may not breed as long as younger males. The decrease in breeding time can result from arthritis, backache, or something physical because of age. We have not experienced any problems with the older males. However, we are aware as they get older the normal effects of aging may interfere with their productivity.

Behavior/ Stress ~ Males

Clueless young males

Some young males might not have a clue what to do with a female. They may need help getting started.

One way to help a male start breeding is to find a female that will cush easily for him. Walk the male behind the female and start making orgling noises so he knows what he is there for. If he still doesn't know what to do, bring an older male into the pen to show the young male what to do. Usually as soon as the younger male hears the older male orgle (sing), they understand what they are there for. The young male will usually start orgling and wanting to breed so quickly remove the older male from the pen.

Some young males may be shy and feel odd having someone lead him around with a halter. Take a step back, leave the young couple together in a small pen, and let nature take its course. Occasionally this approach works well for us.

Young male aggression

When a female is afraid of the male, her body language can appear as a threat to a young male. He might not understand her behavior and he may try to attack her.

A frightened female will, at times, turn sideways to make herself look larger. She may open and close her mouth quickly, put her nose in the air, and lift her tail. To a young male the female's behavior appears as aggression. He may try to attack her. Separate them and try breeding the young female again at another time.

After Shearing

Every year we shear around our spring breeding season. One year when we brought our older experienced herd sire to our fully shorn females, he acted as if the females were intruders. He could not understand what we did with his girls. Every time we brought him to the females to breed, he would try to attack them.

The older male was our only breeding male at that time and his pen was next to the female's pen. We discovered later he was becoming dominant over his territory and it may be why he became aggressive when he didn't recognize his females.

He finally settled down when we had an older female visit our ranch for breeding. At first, he tried to attack her but she ignored him. We kept him on a halter and lead so he couldn't hurt her. She remained in a cushed position and waited for him to breed her. The light finally came on and our herd sire quietly bred her. After breeding the visiting female, he went on to breed his girls without any further problems.

Males line up at the fence to see the females

Note:

To help keep male aggression to a minimum it is best to put some distance between female and male's pens. Housing the males next to the females can be stressful. The males pace the fence and fight for position at the fence line. Some males will become dominant over their territory and become aggressive.

After the incident with our older male, we now have a barnyard separating the pens. A little distance between the males and female's pens enables the herd to remain calmer and healthier.

Flehmen

A male alpaca's behavior when he throws his head back with his nose in the air after sniffing a dung pile used by open females is called Flehmen. Sniffing an open female will cause this strange behavior also.

Flehmen occurs not only in alpacas but in other animals as well. When horses and goats display this behavior, they're funny to watch because their top lip will curl back and it looks as if they are smiling.

Injury/ Infection ~ Males

Fighting teeth

Around age two the male's fighting teeth come in, a sign the male is maturing. When the fighting teeth are present, the male is most likely mature enough to breed successfully.

Male alpacas love to wrestle and they can easily cut one another accidentally with their fighting teeth. Avoid injuries by cutting back their fighting teeth between 2 to 3 years old.

It does not hurt the alpaca to cut their fighting teeth. We cut them back as close to the gum line as possible. Our intact males are together in one large pen so it is important to keep their fighting teeth cut back.

How to cut fighting teeth

Originally, we cut alpaca's teeth with obstetric (ob) wire. In addition, we use a Dremel tool or file to smooth down rough areas on the teeth. The fighting teeth curve backwards and they are located at mid mouth on either side. Alpaca's have six fighting teeth, four on top and two on bottom.

It is difficult to reach the fighting teeth and keep the mouth open at the same time. We use a piece of PVC pipe wrapped with duct tape to hold the mouth open while we cut the teeth. Place the pipe at the back of the alpaca's mouth. The pipe will stick out on either side of their mouth. You will need two people, one to hold the pipe in place and the other to cut the tooth. The fighting teeth are razor sharp so be sure to keep your fingers on the pipe and not in the alpaca's mouth.

We now use an impressive tool that cuts the fighting teeth off quickly and easily. The Fight-O-Matic is a great tool for cutting the fighting teeth off in seconds. Additional information on the Fight-O-Matic can be found under Resources.

Note:

Some veterinarians use clippers to cut the fighting teeth. Do not use clippers because they can crush and fragment the tooth.

Also note, the male's fighting teeth can grow back so be sure to check them once a year. We check the teeth every year when we shear. Shearing time is the best time for us to check teeth because they are in restraints and we have help if we need it.

Injury

Shoulder or back injury

Injuries to the shoulder or back can make breeding uncomfortable for the male. He might only breed for a short time depending on the injury. The decreased breeding time may reduce his chances of getting the female pregnant.

We had a young herd sire injure his shoulder while wrestling with two other males. My husband saw the young male flip over and hurt his shoulder. The young sire could not breed for over a year because of the injury. Eventually he was able to breed but he could not breed more than a few minutes at a time.

Testicular trauma

Trauma to the testicles can cause a decrease in the sperm count. The amount the sperm decreases depends on the extent of the injury. Fewer sperm means less productivity.

Infection

Any infection in the body can cause a male alpaca to be temporarily sterile. The increase in the alpaca's body temperature will decrease the sperm count.

Two males wrestle and show off for the females

Sperm ~ Males

Abnormal sperm

Several years ago, we had a male that could not get the females pregnant. We brought him to Oklahoma State University for an evaluation. They consulted with the top Camelid veterinarian in the country, Dr. Tibary at Washington State University.

After looking at our male's sperm, the doctors stated 99% of his sperm were abnormal and he would never get a female pregnant. The veterinarian's advice was to sell him as a pet and get a new herd sire. It was unfortunate because he was a very nice male. We took the veterinarian's advice and we replaced him and sold him as a non-breeder.

Note:

Oklahoma State University has a teaching hospital and they are a wealth of information. The doctors and teachers are very knowledgeable so every visit to OSU is a wonderful learning experience. Below is a picture of a typical alpaca semen sample.

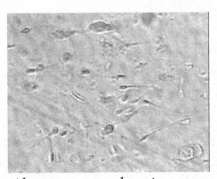

Alpaca semen under microscope

Alpaca sperm sample

An alpaca's semen sample is different looking from other livestock's semen. If the veterinarian never evaluated a male alpaca's sperm, he might mistakenly report the male has a low sperm count and decreased motility. Do not compare alpaca's sperm to a cow or horse's sperm. In comparison, the alpaca's sperm will appear sluggish and lower in sperm count. Find a veterinarian with experience in alpaca reproduction.

Simple semen exam

We check our own males by taking a semen sample from the female's vulva after breeding. We place the slide under a microscope so we can see the sperm. The sperm is checked for abnormalities. When a good sample is obtained, we can see if the sperm have good motility. Obviously, we will not get an accurate estimate on quantity of sperm but at least we can check the quality. Checking the sperm under the microscope is helpful when using a young male to see if he is ready to breed.

Congenital disorders

➢ Hypoplasia of the testicles – smaller than normal testicles results in decreased sperm

➢ Cryptorchidism – undescended testicle results in decreased sperm

Note:

Testicle size does matter. Males with larger testicles can breed more often and more successfully than males with smaller testicles.

Weather ~ Males

Weather too hot or odd weather patterns can cause problems. The following is a few weather related problems we experienced.

Heat Stress

Males may become temporarily sterile because of prolonged exposure to high temperatures. We believe the heat is one of the biggest factors causing sterility in males. Sterility due to extreme heat can last up to 90 days or longer and may even cause permanent damage.

Testicular degeneration can also result from prolonged exposure to high temperatures. The testicles become smaller and the result is a reduced sperm count. If there is too much damage to the testicles, the male alpaca can become permanently sterile.

Summer heat affects fall breeding

Prolonged excessive heat during the summer can destroy sperm. The average cycle of spermatogenesis is 60 days. The 60-day cycle for sperm growth is important to note.

Our fall breeding season starts on October 1st. The first few weeks of breeding we noticed not many females became pregnant. Since spermatogenesis is around 60 days, we can count 60 days backwards and find we are using sperm from August. August is the hottest month in Oklahoma. The year we had severe heat we had difficulty getting the females pregnant when we first started to breed in the fall. The boys were working with less sperm. By waiting a couple of weeks, the females became pregnant and the boys were productive again.

The extreme heat caused one of our males to have severe testicular swelling. We placed him in a room with an air-conditioner and we applied ice packs to his scrotum. It appears since we acted quickly, he did not suffer any permanent damage.

Alpaca wading in a water bucket

Some alpacas love to stand in the water buckets. One way to keep them out of the buckets is to put a few buckets higher out of their reach. We leave one bucket for the feet so they don't get too discouraged and the other buckets are higher to keep the drinking water clean. Everybody is happy, they have clean water to drink while their feet stay cool.

We stop breeding when the weather gets too hot. As summer approaches we only breed early in the morning, while it is still cool, to extend our breeding season a little longer.

The following ideas will help keep the alpacas cool through the summer heat.

Beat the heat

➤ Shear alpacas every spring.

➤ Use large fans to have good air circulation.

➤ Spray the alpacas with cool water on their legs, belly, armpits, and under their tail around the scrotum.

➤ Have plenty of shade available.

➤ Fresh water with electrolytes available at all times.

➤ Rake up excess hay and spray the ground with water. It gives the alpacas a cool place to lie down.

➤ Freeze plastic gallon jugs with water and place them in their water, on the ground, and around their fans.

~ Please remember to shear your alpacas every spring! ~

Two show alpacas after shearing

4 - Nutrition

Supplement

Dry lot conditions require supplementation for the alpacas. Our main alpaca pens are down to dirt and rock because of drought conditions in Oklahoma. Alpacas living in a dry lot condition can become deficient in vitamins and nutrients. If your alpaca pens lack green grass, supplements may be needed. Be aware nutrition can play an important role in the alpaca's ability to reproduce.

Offer around an 11% protein of quality hay for their main staple. We use unfertilized mixed hay with Bermuda and Bahia grasses preferably from the second cut of the season. Supplement with a feed every morning and evening to ensure they receive the vitamins and nutrients they need. We keep loose minerals in every pen as well as a trace mineral salt block. During stressful weather patterns, we also include electrolytes in their water.

Note:

Offer a sheep mineral supplement daily for the alpacas. Both the alpacas and the sheep need copper but they both require copper in small amounts, around 25 ppm when offering it daily.

Vitamin E deficiency

A vitamin E deficiency might exist if the alpacas do not have access to fresh green grass. Over the years, we noticed the alpacas in the pens with green grass became pregnant easier than the alpacas kept in the pens without grass.

Our solution is to keep the areas we have with green grass for our breeding females. A day in an area where they can eat green grass can make a big difference. Rotating pastures helps ensure they will always have green grass to eat.

Selenium deficiency

Selenium deficiency is now a well-known deficiency that can affect reproduction and birthing. The breeding female will be weak and she may have trouble getting pregnant. There are other problems associated with a selenium deficiency such as abortions, stillbirths, and retained placenta. Not only can it affect the breeding female but it can also affect the newborn cria. A selenium deficiency in a cria causes weakness, poor growth, diarrhea, white muscle disease, and even death.

If you are in an area where selenium is deficient, you may need to supplement with selenium injections. Consult with a veterinarian first before giving injections. Test the alpaca's blood levels for selenium because selenium can be toxic if the alpaca receives too much. Therefore an alpaca in a selenium rich area should not receive selenium injections.

To check the amount of selenium by county in the USA visit http://tin.er.usgs.gov/geochem/doc/averages/se/usa.html

Sienna enjoys eating hay

Vitamin D deficiency

The alpaca's skin produces vitamin D when exposed to sunlight. We are breeding for denser fleeces so the sun may not be penetrating through the dense fleece to their skin as well as it should. I never worried about vitamin D deficiencies in our alpacas since Oklahoma is sunny most of the time.

Our feed has vitamin D3 in it yet we still had a few young growing crias deficient in vitamin D. We now routinely include vitamin A & D injections in our young growing crias at 3 months and 6 months of age.

Lack of vitamin D, if not corrected, can lead to Rickets in young alpacas. Osteomalacia is a vitamin D deficiency in older alpacas. The bones become soft and weak in both the young and old. Vitamin D also helps the alpaca's body control the calcium and phosphate levels.

The symptoms of vitamin D deficiency include decreased muscle tone, weakness, bone pain, slow growth, crooked legs, and humped back. They have slower gait and take smaller steps as they walk. With vitamin D deficiency, the female may not be receptive to breeding. If she does become pregnant, she might not carry full term.

Note:

Alpacas will naturally sun bathe. Some alpacas turn their underside to face the sun even when the thermometer is reading over 100 degrees. My guess is they are trying to soak up some of the vitamin D from the sun.

GM Foods

"A **genetically modified organism (GMO)** is an organism whose genetic material has been altered using genetic engineering techniques. GMO's are the source of genetically modified foods..." (Wikipedia online)

We experienced a couple of years where we had a decrease in birth weights and our crias were growing at a slower rate than usual. The feed we were giving the alpacas back then was corn based. After reading many articles online about GM food, we are now thinking the genetically modified corn in the feed might have played a bigger role than we originally thought.

Possible problems related to GM foods

 ➢ Decreased fertility
 ➢ Decreased birth weight of offspring
 ➢ Increased newborn mortality

We read online there are possible issues of infertility in livestock fed GMO foods. There is a possibility these modified foods like corn, soy, sugar beets, and more could be causing an infertility trend in livestock.

Our alpaca feed no longer contains any corn. We are trying to feed healthier organic food to the alpacas, our dogs, and for us too. In order to obtain maximum health benefits, we try to eat our food as close to the way God made it.

Note:

For more information on GM foods and the health risks in livestock and humans, visit the following sites:

http://www.responsibletechnology.org/health-risks

http://www.organicconsumers.org/articles/article_11361.cfm

http://www.actionbioscience.org/biotech/pusztai.html

5 - Breeding

Breeding Methods

Pasture Breeding

During pasture breeding or field breeding the male alpaca remains in the pen with the females. We know several breeders that pasture breed their alpacas successfully.

Since alpacas are induced ovulators leave the male with the females only a short time. If a female aborts or has fetal resorption, the male will breed the females as they become open. When the male alpaca remains in the same pen with the females all year, there is also a risk the crias will be born at odd times during the year.

One breeder we spoke with said she wraps her female's tails with Vetrap. The male remains in the pen with the females for one month. After one month, she removes the male from the female's pen. She waits one week and spit tests the females. If a female is open, she has the male breed the open female. She retests the females with the male in one week to make sure they are all pregnant.

The main drawback in pasture breeding is the breeder does not know the exact breeding dates. Occasionally she was lucky and saw a male and female breeding but most of the time she only had due dates down to one month.

Two females interested in the breeding pair

Note:

When pasture breeding, put only one male with a small group of breeding females. By keeping one breeding male with the females, confusion with parentage can be avoided. Do not put females too young to breed in a pen with the breeding male. The male will breed all the females including the ones that are too young to breed. As previously stated it is best to remove the breeding male from the female's pen at least two months before the females give birth.

Hand Breeding

When hand breeding, the male and females live in separate pens and they are only brought together for the breeding. At our ranch, we put a halter on the male and bring the male to the female's pen to breed. We prefer hand breeding because we can witness and record every breeding that takes place on our farm.

One advantage of hand breeding is being there to take care of any problems that need our attention. After breeding, we check the female's vulva for discharge. If we need to get a sperm sample, we take it from the female's vulva after breeding. We take note how long the male breeds the female. We also note if there is any interruption during the breeding because some males will get up and down when they breed.

At the end of a breeding season when most females are pregnant, hand breeding takes a turn. The females spit when they are pregnant so make sure to wear an old shirt because the spit will fly towards the male and the person leading the male. Spitting females can be overwhelming so test only a few females at one time.

Since we know our exact breeding dates, we can calculate our due dates for our crias. We calculate the female alpaca's due date by a 340-day gestation. However, as you know gestation can be anywhere from 330 to 365 days or longer. Visit our cria due date calculator at www.walnutcreekalpacas.com/cria_calculator.htm to see when your female is due!

Breeding Frequency

Females

How often can a female alpaca breed?

Only breed a female alpaca once a week. If the female cushes before the week is up, continue to wait the full seven days before breeding her again. We have much less incidence of uterine infection by waiting seven days between breedings.

Young male cria practicing breeding

How soon can a female breed after having a baby?

Breed the healthy female back to the male, after she delivers her cria, no sooner than 17 days post partum. If the female has complications from the birthing process, such as a dystocia, wait at least 30 days post partum before breeding her back to the male. Sometimes we wait longer depending on the complication to give her body a chance to heal. We skip an entire breeding season or two if the female has a prolapse or c-section.

Males

How often can a male alpaca breed?

A breeding male's age determines how often he will breed. We breed our experienced males once a day. Our young males just starting out will breed every other day or every three days depending on the male and his age.

Overusing a young breeding male

One season we were breeding one of our young males every day. I read male alpacas can breed up to 4 times or more in one day so I thought we could use this male every day with no problem. There was a problem because only 1/3 of the females became pregnant. We called Oklahoma State University and they said we were overusing him. They advised backing off the breedings and use him only every other day. Well that was excellent advice. Once we decreased his breedings to every other day, all of our females became pregnant that breeding season.

Note:

We would rather spread the breedings out than risk the males being less effective or start shooting blanks. We want our males to impregnate our females with as few breedings possible.

Location/ Strategies

We experiment and try different locations to improve the breeding process so it will run smooth and efficient. Every alpaca breeder may run his or her farm differently. Do what works best for your situation. The following information includes a few ideas that work well for us.

Several females cush by a breeding pair in the barn

Some males are dominant over their territory. It is best to bring the male alpaca to the female and not the other way around. When we bring the male to the female we keep the halter and lead on the male at all times. We never leave the pen so we are present for the entire breeding.

Breeding a young female

When we have a young female afraid of males, we put her with 2 or 3 calm females in a small 10 x 20 area and bring the male to them. The young female should cush with the other calm females present. If she will not cush, bring a different male in the pen to breed one of the other females. Once the young female relaxes and cushes, bring her mate back in the pen to breed her while the other breeding is going on. Breeding pairs side by side helps ease the tension for young females. We have successfully bred several frightened maidens this way.

Note:

When breeding alpacas find an area that has few distractions. We don't mind the distractions from the alpacas themselves as long as it is minimal. We try to avoid outside and unfamiliar distractions. It is best to reduce the stress level and keep everything calm.

Breeding a young male

When using a young male find an area that is far enough away from the adult male's pen. Adult males can be intimidating to a young male starting to breed. If the young male can see the adult males, he may be too frightened to breed.

Moon Phase Breeding

Oops wrong way!

How to breed by the moon

First things first, what is moon phase breeding? Other alpaca breeders explained to us if we breed our alpacas on the waxing side of the moon, we would get more females. The waxing side of the moon is from the new moon to the full moon. Breeding on the waning side of the moon produces more males. The waning side of the moon is from the full moon to the new moon.

To help make breeding by the moon easier there are calendars available that show what stage the moon is in each month.

We heard about moon phase breeding from several breeders over the years. We never gave it much thought but we finally decided to check our past breedings to see if moon breeding had any possible merit.

We laughed about moon phase breeding until we checked our breedings, we found 90% of our breedings coincide with the phase of the moon. The girls were coming from breedings on the waxing side of the moon and the boys were coming from the waning side. More female alpacas... we thought we were on to something!

What we found out was so incredible. We started to breed by the moon's phase and initially we did have more girls. However as we continued to breed by the moon, the percentage went down and little by little we were having more boys.

Well God showed us who is in control. I guess there has to be balance in nature... the way God designed it. This past year our boy to girl ratio continues to run about 50/50.

6 - Herd Sire Selection

When choosing a male we obviously want to pick the best male we can find for our females. However, there are several things to consider before pairing up the males with the females next breeding season.

Breeders should put together a breeding plan for their herd. Decide what is important and what goals to strive for. We know every breeding program is going to be different but we feel it is important to improve the herd with every breeding if possible.

What to look for in a breeding male:
> Registered pedigree
> Perfect conformation and bite
> Fine fleece
> Uniformity of fleece
> Dense fleece
> Abundance of fleece character such as crimp, brightness, and lock structure
> Easygoing disposition

We also want the male to hold the wonderful fleece qualities we mentioned above as he ages. Find a male that will improve the female; a male strong in the areas the female is weak in.

Breeding decisions can be as simple as breeding a female with coarse fleece to a male with fine fleece. Breeding decisions become difficult when the female's needed improvements are small and refined.

It is important to learn about different bloodlines. Some of the top bloodlines in the country are producing outstanding offspring across-the-board for a reason. They are proven over the years to produce top quality crias.

Some farms are breeding for specific colors or for fleece with outstanding character. It is important to plan and decide what you want for your farm. Only you can answer those questions. Always have goals and a plan to attain your goals.

Taking a chance on a young male

Have you thought about taking a chance on a young male? There are some things you might want to consider. A young male could have everything you want in a breeding male but as he grows, he could lose everything that drew you to him. On the other hand, you might get lucky and find a male that improves with age. The next hurdle is when it comes time to breed. There is no way to know if the male will throw average or outstanding crias.

When buying a young male you are essentially taking a chance. However, there are certain qualities to look for when choosing the right male for your females.

The young male we took a chance on

Everyone we meet at the alpaca shows asks about our top herd sire, Classic Peruvian Magic. The big question... how did we know he would turn out so great and an 8x champion to boot!

We first saw Magic as a newborn cria for sale on the Internet. He is from proven bloodlines and he has an excellent pedigree. His picture revealed little but we could see there was something special about him.

Magic at 3 months

When we were finally able to see him up close and put our hands on his fleece, we knew we had to have him. His fleece was extremely fine and full of character. His conformation was impeccable and he had the presence of a true herd sire at only three months of age. We purchased Magic when he was three months old. Prayers were answered and we were blessed. As Magic grew, he continued to improve every year.

Magic at 1½ years old

Magic's fleece

Magic's fleece histograms are impressive. His first histogram in 2007 revealed a very fine fleece with a 14.7 average fiber diameter. In 2009, his histogram was 17.7 average fiber diameter. In 2010, his shearing weight was 10.7 pounds of very fine fleece.

At 4 years old at the MOPACA show, the judge stated Magic was the oldest and the finest of the brown alpacas in all the brown classes at the show including the crias. Magic won a blue ribbon and brown championship at that show.

We didn't fully realize his potential as a herd sire until his offspring arrived. He is passing so much of his superior traits on to his crias. He is prepotent and his offspring are outstanding. There is some guesswork about a herd sire's genotype by looking at his bloodlines. However, the true genotype of an alpaca is not fully apparent until they have offspring on the ground.

How did we know Magic would turn out to be a fantastic choice for herd sire? We pray over every alpaca we purchase and we leave everything in God's hands. Magic turned out to be a wonderful blessing for us.

Phenotype versus Genotype

Phenotype and genotype are terms used by breeders. When a breeder uses the term phenotype, they are talking about an alpaca's appearance. Genotype on the other hand is the alpaca's genetic makeup so his pedigree is very important.

When finding the right male for your females, look at the alpaca's phenotype but the genotype is also important. In other words, find a male that will back up his good looks and pass those excellent qualities on to his offspring.

Linebreeding

When new alpaca breeders visit our farm, the question about linebreeding usually comes up in the conversation. We live in a rural area where there are many farms. Linebreeding is done in other livestock so they want to know if they can line breed alpacas. Alpacas are no different than other livestock and yes if done properly linebreeding can be done.

Breeding in the female's pen
is a good learning experience for the young crias.

We prefer not to linebreed. I am not a geneticist and everyone I meet has different opinions about linebreeding so I'm not going to go into any details here. If you have questions about genetics or need more information on linebreeding please refer to Resources, Alpaca Genetics Library Online.

If you need additional information on alpaca pedigrees, the Alpaca Registry (ARI) has a superior database. The Alpaca Registry's information is listed in the Resources section.

Breeding for color

Predicting the color two alpacas will pass to their offspring is complex and often difficult to predict. I will keep it simple and share a few basics we learned over the years. The following results are what we observed.

Results of breeding certain colors:

➢ A white alpaca will throw white offspring about 50% of the time.

➢ An alpaca with a white spot will throw about 50% white spot pattern and 50% solid color.

➢ Vicuna (fawn) alpacas usually throw the color they are bred to.

Color in general

When breeding alpacas it becomes apparent that white is a dominant color and black is a recessive color. For years we bred a black sire to our females to find out what other color our girls would throw. Recently I discovered an online article by Andrew Merriwether titled "Determining the Recessive Color" and now our using black to learn the secondary color makes sense. His article is very informative plus it is easy to understand. For more information on color and genetics, please refer to the Resources section, Alpaca Genetics Library Online.

White spot

When breeding two white spot alpacas together there is a 25% chance the offspring will be white with blue eyes. Sometimes deafness is associated with the blue-eyed whites so we try not to breed two white spot alpacas together.

How to avoid blue-eyed white offspring

A general rule would be not to breed two white spot alpacas together. Also do not breed a white spot alpaca to a white if you do not know if the white alpaca has a white spot or not.

We can breed any color including females with a white spot to our white herd sire. AJ's Peruvian Alliance is white but he does not carry the white spot gene so he can never throw a blue-eyed white. He has proven over the years he does not possess the white spot gene so he can safely breed females with a white spot.

Rossi is an example of a Tuxedo grey

The tuxedo greys usually have a white face with some white on the neck and legs. Tuxedo greys carry the whitespot gene.

Grey alpacas

When breeding greys we prefer to breed a fawn (vicuna) to grey to improve the fleece quality and this has worked well for us over the years. Walnut Creek Rossi is the outcome of such a breeding. Rossi is out of Masterpiece Ruth, a tuxedo medium silver grey female and he is sired by BF Giacomo, fawn with vicuna pattern. Giacomo, being vicuna colored, will usually throw the mother's color plus he also covers black. He was the perfect choice for Ruth to have a superior silver grey and she did! Below is Rossi's impressive superior fleece.

Rossi's superior grey fleece

Before breeding tuxedo grey alpacas, be sure to investigate all possible results. After talking with several breeders on breeding grey to grey, it is apparent there is important information you need to be aware of, such as blue-eyed whites and the lethal grey. If you are serious about breeding greys and need more information visit http://www.grayalpacacentral.com they have helpful articles on the genetics behind breeding grey alpacas.

7 - Proven Breeding Schedule

Day 1

Bring the male to the female. Keep the male haltered and lead him to the female.

> If the female spits or runs away from the male - bring the male back to her in a couple of days.

> If the female cushes - the male will breed her. Wrap the female's tail beforehand or simply move her tail out of the way. Make sure the male penetrates the female to ensure a successful breeding. Also check to ensure the male does not get his penis wrapped up in the female's tail.

Note:

The breeding can last anywhere from 5 minutes to an hour. Most alpaca males will breed around 20 minutes.

Day 7

Bring the male back to the female.

> If the female spits or runs away from the male – this means she has ovulated

> If the female cushes – go back to day 1 and start over.

Note:

Studies on the frequency of breeding alpacas show that breeding more than once every seven days does not improve the chance of getting the female pregnant. Increasing the frequency of breeding only increases the chance of injury or infection. Please remember in this case, more is not better.

Day 14

Bring the male back to the female.

> ➢ If the female spits or runs away – this means a corpus luteum formed and the egg may be fertilized.

> ➢ If the female cushes – go back to day 1 and start over.

Day 21

Bring the male back to the female.

> ➢ If the female spits or runs away from the male – the egg implants and most likely she is pregnant!

> ➢ If the female cushes – go back to day 1 and start over.

Note:

Do a progesterone test around day 28 if you feel one needs to be done. It is not necessary to do the progesterone test at this time. When using different labs to do progesterone testing find out what your lab considers pregnant... (more on progesterone testing in chapter 8)

Day 30

Bring the male back to the female and do a spit test

Day 40

We use the Preg-Tone to confirm pregnancy... (more on using the Preg-Tone in chapter 8)

Note:

After 3 or 4 breeding attempts in one breeding season, if the female is not pregnant, further investigation is needed.

Different Scenarios

The female will not cush:

➢ The female might be pregnant

➢ She may have a retained Corpus Luteum (CL)

Tests – progesterone test and ultrasound

The female continues to cush every 7 days:

➢ Age related - she may be too immature

➢ Possible uterine infection – check for discharge

➢ The female may not be cycling or ovulating

Tests – check for uterine infection and ultrasound

The female continues to cush every 30 days or more:

➢ Check progesterone levels - she might have a progesterone deficiency and unable to maintain pregnancy.

➢ She could have a uterine infection – there may or may not be discharge.

Tests – progesterone test, check for infection, and ultrasound

Reproductive Examination

If your female alpaca has not been able to conceive contact your veterinarian and have a full reproductive examination done. Be sure to give your veterinarian complete and accurate medical and breeding records.

Possible procedures the veterinarian will perform:

> ➢ Check for discharge
> ➢ Vaginal examination
> ➢ Check for physical abnormalities
> ➢ Lab work – draw blood and take a culture to rule out infection
> ➢ Check progesterone levels
> ➢ Ultrasound examination
> ➢ Check uterus and ovaries
> ➢ Check for retained Corpus Luteum or any other abnormalities

Note:

Do not get discouraged, our veterinarians could not find anything wrong with a few of our females even after a full reproductive examination. By waiting a breeding season or two, some of the females went on to breed and produce beautiful healthy crias.

Your alpaca's breeding records are very important. Your records combined with a thorough examination will help the veterinarian find out why your female is not getting pregnant.

Possible treatment plan:

> ➢ A uterine flush is helpful to cleanse the female's uterus.
>
> ➢ Antibiotics are given if an infection is present.
>
> ➢ Hormone treatment can be helpful; Estrumate helps the female to start cycling.
>
> ➢ Other hormones used such as Cystorelin can help the female ovulate.

A male tests the female to see if she will cush

8 - Pregnancy Testing

Alpacas have a higher rate of early pregnancy loss than most livestock so it is important to test your females. To ensure pregnancy we test some of our females after 40 days and then again at 120 days.

Behavior or Spit test

Spit testing can be accurate if you know the female's behavior in advance. Each female will react to the male a little differently. Some females are easy to read and others are not so easy.

Ideal Basic Response
 ➢ Pregnant – Bring the male to the female and if she is pregnant she will spit at the male.
 ➢ Not Pregnant – An open female will cush for the male and she will allow the male to breed her.

Note:
Not all females spit when they are pregnant or cush when they are open. If every female would follow the ideal basic response, life would be so much easier. Unfortunately, most of our females give different responses and this is why it is helpful to know the female's behavior in advance.

Walk-through breeding check

Several times a year, we bring a male into the female's main pen for a walk-through breeding check. We keep the male on halter and lead at all times. As he is led into the female's pen he usually orgles (sings) and will continue to sing the entire time he is in their pen.

Some females will cush and others will spit and run away from the male. Be aware you might be spit on when doing a breeding check this way. Be sure to wear a raincoat if there are several spitting bred females in the pen!

Male and female interacting over the fence

Note:

Keep your eyes on the male at all times. Make sure he doesn't jump on the young crias as you do the walk-through check.

Preg-Tone

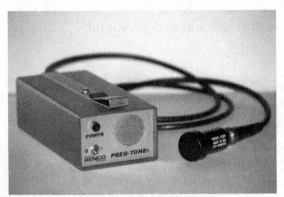

Preg-Tone ultrasound device

The Preg-Tone is an ultrasound device using sound rather than an image to confirm pregnancy in animals. The Preg-Tone is the most reliable and the easiest way for us to test our females with high accuracy.

Using the Preg-Tone

The Preg-Tone is easy to use but it takes two people, one to hold the alpaca and the other to operate the Preg-Tone.

The Preg-Tone works well any time after 40 days after conception. It can be used earlier but pregnancy is much more difficult to detect earlier than 40 days. We prefer to wait until 60 days after conception.

Clip the Preg-tone to your belt or pocket. Generously rub mineral oil on the alpaca's belly and on the ultrasound head. If the alpaca is very hairy on her belly, use a generous amount of oil and smooth the hair down to get rid of air bubbles. If the ultrasound head has good contact, the unit will make a beeping noise. If the ultrasound head does not have good contact, the unit will not make a sound.

Point the ultrasound head in the direction of the arrow

We usually start on the left side first since most alpaca's pregnancies occur in the left uterine horn. If there is no tone, we move to the right side. If the female is pregnant, the Preg-Tone will emit a continuous tone.

The Preg-Tone from our experience is approximately 99% accurate. It is an extremely valuable tool in detecting pregnancies on our ranch.

Instructions for alpacas from Renco's site

"Testing can begin at 30-40 days. Alpacas and llamas are tested on the left side. If unsuccessful, try the procedure on the right side. Place probe 6-8-in (15-20cm) down from the udder. (On maidens about 5-in (12cm)). Point probe straight into the abdomen at an angle defined by the shape of the abdomen."

Note:

See the picture on the previous page... the arrow on the alpaca shows the direction of the probe's angle. Usually two people are required to test a female with the Preg-tone.

Progesterone test

Progesterone tests are not very reliable since the alpaca's progesterone levels will fluctuate during pregnancy. Different labs show different values for negative and positive. Usually progesterone levels less than 1 ng/ml indicates not pregnant, 1-2 ng/ml means their not sure, and greater than 2 ng/ml is positive for pregnancy. Please note progesterone levels will also be elevated when the female has a retained Corpus Luteum (CL).

Note:

We did a progesterone test on a pregnant female and the test came back negative, it was < 1 ng/ml. We spit test the female and she spit at the male. We knew she was pregnant by the way her personality changed. Another progesterone test revealed she might be pregnant. The female went on to have a healthy cria the following spring. We did not have the Preg-Tone at that time otherwise the Preg-Tone would help confirm her pregnancy easily.

Ultrasound

An experienced veterinarian must do the ultrasound exam. Please note, if the veterinarian doesn't see anything, it does not mean she isn't pregnant.

A word of caution; we hear too many horror stories of farms doing an ultrasound on a female and the veterinarian states their female is not pregnant. The veterinarian gives the female Estrumate (cloprostenol sodium) to get her to start cycling but instead of cycling the female aborts a fetus.

Note:

If the female is too far along in her pregnancy, the ultrasound test may not show a fetus. To understand why this happens... compare it to standing too close to something so large that you can no longer identify what it is.

Magic's first cria, a few hours old

9 - Additional Information

Transporting in Last Trimester

Do not transport a female in the last trimester of pregnancy. The female alpaca's colostrum, first milk, contains antibodies to local bacteria and viruses where the alpaca lives. When moving a female in her last trimester her body does not have time to make antibodies to the new farm's bugs before she delivers. If the female does not have adequate antibodies to the new bacteria, she will be unable to pass them to her newborn cria. The cria usually dies within the first 3 days of life.

Our ranch as well as other ranches experienced the loss of crias because of the lack of local antibodies after transporting females in their last trimester. We had two incidents where the pregnant females did not have the time to make the antibodies needed for their crias.

First Incident

The first incident occurred when we transported four bred females from Oregon to our farm in Oklahoma. The females arrived about 3 months before their babies were due. The first cria born died within a few hours after birth. The second female delivered her baby a few days later and her cria only lasted a few hours longer. The cria died on the way to OSU. The third female's cria lasted longer. She did well during the day but by evening, the cria became septic and we rushed her to OSU. I prayed the entire way to OSU and asked God for a miracle. When we arrived at OSU, they asked for her name and I said, "Her name is Miracle!" because I knew she needed a miracle to live. God is so good and He continues to answer prayers. Miracle is still with us today... happy, healthy, and making beautiful crias for us.

By the time the fourth female was due I was emotionally exhausted and I knew I could not go through another cria battling for life. We brought the female to OSU so she could have her baby there. Her newborn cria became septic only a few hours after birth. Since she was born at OSU, they were able to treat her immediately and they saved her life. It was apparent the Oregon females lacked the antibodies to the germs in Oklahoma. Shortly after our ordeal, OSU had to help another ranch with the same problem.

Second Incident

The second incident occurred when we sent two pregnant females, both due within a couple of weeks, to a farm only 20 minutes away. We thought the newborn crias would do fine at the other ranch because they were so close. Surely, they would have similar bacteria and germs to our ranch. Unfortunately, we were wrong and both crias died within a couple of days.

Note:

Transporting distance does not make a difference so be cautious and do not transport a pregnant female in her last trimester.

Please take note:

Colostrum supplements are not created equal

Some colostrum supplements do not contain any antibodies! Obviously fresh colostrum would be ideal and frozen would be second best on our list. There are times when the fresher types are not available and a supplement is needed.

When using a colostrum supplement make sure it contains antibodies and it is a colostrum replacement not just a supplement.

Resources

Alpaca Owners & Breeders Association Inc. (AOBA)
5000 Linbar Drive, Suite 297
Nashville, TN 37211
Phone: 615-834-4195, USA only: 800-213-9522
http://www.alpacainfo.com

Alpaca Registry Inc. (ARI)
8300 Cody Drive, Suite A
Lincoln, NE 68512
402-437-8484
http://www.alpacaregistry.com

Alpaca Genetics Library Online
The G.A.I.N.
Library of genetic articles written by Dr. Wayne Jarvis,
D. Andrew Merriwether, Ph.D., and Ann M. Merriwether, Ph.D.
http://www.alpacagenetics.com/library.html

Australian Alpaca Association (AAA)
Unit 2/613 Whitehorse Road
Mitcham, Vic 3132
Telephone: +61 (0)3 9873 7700
http://www.alpaca.asn.au/

Canadian Llama & Alpaca Association (CLAA)
2320 41 AV NE
Calgary, AB T2E 6W8
Phone: 403.250.2165
Toll Free: 1.800.717.5262
http://www.claacanada.com

Breeding and record keeping
HerdEase / AlpacaEASE
Alpaca Management Software
by Ellen Gurewitz
www.alpacaease.com

Oklahoma State University (OSU)
Large Animal Veterinary Clinic
1 BVMTH
Stillwater, OK 74078
405-744-7000

Pregnancy testing
Preg-Tone
www.rencocorp.com/preg-tone.htm
Instructions from Renco
http://www.rencocorp.com/pdf/pregtone_instructions_2009.pdf
Preg-Tone instructions for use on an alpaca
http://www.rencocorp.com/pdf/AlpacaLlamaSupplimentalInstru
ctions-3.pdf

M& M Veterinary Laboratory - Progesterone testing
Mel Hoskin, M.T.
13615 Wabash Road
Milan, Michigan 48160
734-439-2698

Selenium map
Counties in the United States
http://tin.er.usgs.gov/geochem/doc/averages/se/usa.html

The Fight-O-Matic - Teeth Trimming
The AlpacaRosa
Don & Carolyn Marquette
http://www.thealpacarosa.com/fightomatic.html

The International Lama Registry - Alpaca Division
PO Box 8
11 ½ S. Meridian
Kalispell, MT 59903
(406) 755-3438 voice
https://secure.lamaregistry.com/Beta2/registry-services/alpaca-division.php

Walnut Creek Alpacas - Gestation Calculator
http://www.walnutcreekalpacas.com/cria_calculator.htm
Find out when your female will give birth. A gestation cria calculator based on 340 days.

References & Further Reading

Caring for Llamas and Alpacas
A Health and Management Guide
Clare Hoffman, DVM and Ingrid Asmus
Rocky Mountain Llama and Alpaca Association, 2000

International Camelid Institute (ICI)
Founded by David E. Anderson, DVM, MS, Diplomate ACVS, at
The Ohio State University College of Veterinary Medicine
http://www.icinfo.org/

Key Reproductive Features
Australian Alpaca Association, LTD
Male and female alpaca anatomy and physiology
http://www.alpaca.asn.au/docs/about/info/1reproduction.pdf

Llama and Alpaca Neonatal Care
Bradford B. Smith, DVM, PhD, Karen I. Timm, DVM, PhD,
Patrick O. Long, DVM
Bixby Press, Corvallis, OR, 2001

Llama * Alpaca Field Manual
C. Norman Evans, DVM
Evansville Bindery Inc., Indiana, USA 2001

Medicine and Surgery of South American Camelids
Murray E. Fowler, DVM
Iowa State University Press, Iowa, USA 1989

Reproductive Physiology of South American Camelids
Walter Bravo, DVM, MS, Ph.D.,
Department of Population Health and Reproduction, School of
Veterinary Medicine,
University of California, Davis

The Complete Alpaca Book
Eric Hoffman
Bonny Doon Press, Santa Cruz, CA, 2003

The Postpartum Llama: Fertility after Parturition' 1994
P. Walter Bravo, Murray E. Fowler, and Bill L. Lasley
Departments of Population Health and Reproduction3 and
Medicine, 4 School of Veterinary Medicine, University of
California, Davis, California 95616

Uterine infections in Camelidae 2001
Professor Ahmed Tibary
Department of Veterinary Clinical Sciences, College of
Veterinary Medicine, Washington State University, Pullman WA
99164-6610
Professor Abdelhaq Anouassi
Veterinary Research Centre, PO Box 77749, Abu Dhabi, United
Arab Emirates

Appendix

Glossary

AOBA – Alpaca Owners and Breeders Association

ARI – The Alpaca Registry Incorporated

Conception – the beginning of pregnancy... the union of egg and sperm to form a fetus.

Congenital disorders – a condition that dates from the time of birth. It can be inherited or caused by an environmental factor.

Corpus Luteum (CL) – a small, yellow progesterone-secreting structure that forms on the ovary after an egg has been released.

Cria – a baby alpaca from birth to weaning age.

Cryptorchidism – undescended testicles, failure of one or both of the testes to descend into the scrotum.

Cush – (or Kush) a sternal sitting position with the legs tucked underneath.

Dystocia – a slow or difficult labor or delivery.

Estrous – a regularly recurring period where the female is most receptive.

Fighting teeth – eight curved sharp teeth in mid mouth. Four on top and two on bottom.

Flehmen – a particular type of curling of the upper lip in many mammals, which facilitates the transfer of pheromones.

Follicle – ovarian follicles are composed of roughly spherical aggregations of cells found in the ovary. They contain a single immature ovum or egg.

Freemartin – an infertile female, which has nonfunctioning ovaries. She is sterilized in the womb by hormones from a male twin. Freemartinism is the normal outcome of mixed-sex twins.

Genitalia – externally visible sex organs

Genotype – the genetic makeup of an animal

Gestation – the carrying of an embryo or fetus inside a female

GMO – Genetically Modified Organism - an organism whose genetic material has been altered using genetic engineering techniques. These techniques use DNA molecules from different sources, which are combined into one molecule to create a new set of genes. This DNA is then transferred into an organism, giving it modified genes.

Gonadotropin Releasing Hormone (GnRH) – the hormone responsible for the release of follicle-stimulating hormone (FSH) and luteinizing hormone (LH) from the anterior pituitary.

Hypoplasia – the underdevelopment of a tissue or organ.

Luteinising Hormone (LH) – a hormone produced by the anterior pituitary gland. In females, an acute rise of LH called the LH surge triggers ovulation and development of the corpus luteum

Meningeal worm – also known as a brain worm, carried by the white tail deer

Osteomalacia – a disease in adults characterized by softening of the bones, resulting from a deficiency of vitamin D, calcium, and phosphorus

Ovary – the reproductive organ in female animals that produces eggs and the sex hormones estrogen and progesterone. The ovaries contain numerous follicles, which house the developing eggs.

Ovulation – the release of an egg cell (ovum) from the ovary in female animals, regulated by hormones

Phenotype – the physical appearance of an animal

Pituitary – the master gland of the endocrine system; located at the base of the brain

Postpartum – occurring immediately after birth

Prepotent – effective in transmitting hereditary characteristics to its offspring.

Progesterone – a steroid hormone, secreted mainly by the corpus luteum in the ovary, which prepares and maintains the uterus for pregnancy.

Rickets – a disease of the young caused by deficiency of vitamin D and sunlight associated with impaired metabolism of calcium and phosphorus.

Spermatogenesis – formation and development of spermatozoa.

Uterine prolapse – the slipping or falling out of place of the uterus.

Vulva – the external genital organs of the female.

Index

Peruvian Magic & Calendar Girl

All of the information provided in this book is based on either personal experience or information given to us by veterinarians for the treatment of our alpacas. Please consult your veterinarian before using any information provided by this book. Neither Dr. KD Galbraith or walnutcreekalpacas.com nor any of the contributors to this book will be held responsible for the use of any information contained herein.

About the Author

Dr. KD Galbraith holds a doctorate degree in chiropractic plus degrees in nutrition and physiology. She has over 12 years experience raising and breeding hundreds of alpacas.

In collaboration with her husband, David Galbraith, she has created several successful websites and written several articles. Her passion is alpacas along with photography and writing. You're welcome to visit her online at their alpaca website and blog at www.walnutcreekalpacas.com

CPSIA information can be obtained
at www.ICGtesting.com
Printed in the USA
BVOW03s2232141217
502506BV00005B/319/P